An Introduction to Project Management

Mind the Gap

J. B. Blacklock. B.Sc., C.Eng., MICE.

The moral right of John Barry Blacklock to be identified as the author of this work has been asserted by him in accordance with the Copyright Designs and Patents Act 1988.

Disclaimer: This book contains general information and views that are based on the author's own knowledge and experience. As such the author disclaims any liability, directly or indirectly, based on the application of the contents by others.

Contents

Preface

This book provides experience based information for students and those new to the arena of project management, as an introduction to project management.

Subjects are discussed to provide initial thought provoking consideration on the ideas presented. These subjects are presented in a sequence aimed at building your project management knowledge in a progressive way. As you read into the book the previous chapters become interconnected with the current text. This reflects the nature of managing a project.

An experience based extended definition of goals (from SMART to SMARTER) is described to encourage you to enjoy your work and to reward yourself, and your team, for your achievements.

As everyone is different, the intention of this book is that you will start to acquire some knowledge from the book. You should then feel enabled to go on to further reading plus developing your own experience. As your knowledge accumulates you should go forward and grow into being the successful project manager you wish to be.

J. B. Blacklock.
April 2015.
Market Deeping, England.
baz-b@outlook.com

About the Author

John Barry Blacklock is a Chartered Civil Engineer. He has been a designer of sewage treatment works and their extensions, a Design Team Leader and a Project Manager. For a short time, following qualification, he carried out the role of Life Coach in support of those looking to improve their lives with the support of non-directive coaching.
He has more recently worked on the design of drinking water works.
His experience, on which this book was based, was developed during working in two water companies, two engineering consultancies, and a mechanical engineering company.

Introduction

Welcome to 'An Introduction to Project Management'.

This book has been written for those of you who are looking to open a door and peer into the world of Project Management. The content reflects over 25 years experience on the successful design, leading of design teams and project managing the delivery of civil engineering works in (for the most part) the sewage treatment industry. By scaling the extent and detail of these project management ideas, skills and techniques you can apply them to achieve successful events (holidays, birthday parties and private events).

Let's get started by considering the question 'What is a Project?' For the purposes of this book a Project has been defined as:

A client's goal required to be achieved, through relevant plans and tasks, to a time, to a budget and to a defined quality.

Delivering your project needs people power, to some degree or other. Therefore, it's valuable to identify a few facts about people (and generally that includes Project Managers!).

About people - Did you know that:

- Only 1% of what we see is in focus (so 99% of what we see is not)

- The human brain records at about 3 frames per second (so whatever happens faster than that we don't record)

- A human brain can only carry out a limited number of tasks at a time, and it's believed to function optimally when it works on one task at a time. Therefore, when you are multi-tasking you may not be working optimally. (Does that mean the more you multi-task the less optimal you become?)

- High performing individuals can complete a multiplicity of tasks, one after another (so not at the same time)

- Most healthy adults are unable to sustain attention on one thing for more than about 40 minutes at a time

Equipped with this information about our own capabilities, what other key information may be of value to us as project managers? In delivering many project tasks and events we can benefit from the support of computer power, to some degree or other. Therefore, it's valuable to identify a few facts about computers too.

About Computers – Computers can help us, and they can hinder us, in our days work. To effectively use any tool in achieving our goals it's good to have knowledge about the tool itself.

Some facts from 'Computer System Reliability: Safety and Usability' (Dr. B. S. Dhillon: 2013) [1]:

- In 2002, a study commissioned by the National Institute of Standards and Technology (NIST) reported that software errors cost the U.S. economy about $59 billion annually [2]

- As per Landauer [3], an average software program contains about forty design flaws that impair the ability of workers to use it effectively.

Separately a search of the internet reveals that:

- The Internet is the fastest-growing communications tool ever. It took radio broadcasters 38 years to reach an audience of 50 million, television 13 years, and the Internet just 4 years

- Looking at the NASA website and searching for information regarding computers reveals that when NASA successfully sent the Apollo missions to the moon – "The Apollo on-board computer system was integrated so fully into the spacecraft that designers called it "the fourth crew member". [4]

It can be seen, therefore, that as far back as in the 1970s, NASA identified that they needed computers to get their tasks done. They were putting a rocket on the moon which may be more than you are doing. However, to get numerous tasks to come together to achieve an event or project section, a computer could be as valuable to a project manager, as a team member.

Having defined what a project is, and with a sense of how good people are in the fields of seeing, observing and performing, and the ups and downs of the use of computers, let's explore what we can achieve with these in Project Management...

Goals

As mentioned in the definition of a project, you will have your clients' goals to achieve. These need to be defined by your client, as it's their project. The degree to which your client is able to express and specify their goals will contribute to the robustness of the foundation, and successful delivery, of the project. Spend time with your project owner to find out what really is needed from the project. Agree a definition of what the project should, and should not, deliver. Knowing what a project is, as well as what it is not, provides you with boundaries.

Will these boundaries be clear? Sometimes they will. Sometimes they cannot be – until more becomes known as the project moves forward. A catch 22 situation! Well, it's alright to become aware of and to accept that there will be things you don't know. This *awareness* is a positive aid for you. This is because you can manage these unknowns as part of the project activities, as they move towards becoming *known*.

Sometimes your projects initial boundaries (known or unknown) may need to be changed depending on the progress of the project. This need for change may come about in reflection of information that becomes apparent during the project life cycle. If this wasn't true then you wouldn't be needed. Instead the requirements for the project could be entered into a computer and, as if by magic, the finished project would materialise without problems, issues or risks. It would arrive in time and within budget, a perfect fit to the client's specification.

You are needed precisely because projects need to have all this detail, all the product information and all the stakeholders requirements, met – to the needed degree. Often the definition or specification, for the reasons just outlined, will be lacking to some extent. The scale and impact of this lack of information will vary depending on the client, the project delivery model, the goals and the involved stakeholders. As a project develops, more information develops with it. Much of this needs human intervention, at varying levels, to interpret, develop and define ways forward that fit into, and enable, delivery of the goals. Most likely this intervention will need you (what would the project do without you) to collate the information and, from that information, to propose modification to the goal details, and related time, cost or quality definitions.

You need to know, from this point on, you are likely to find the projects that run smoothly tend to be those which are brief in duration and are a repeat of what has gone before. Much like making a cup of tea – but hang on you've just run out of tea bags ...

Conversely, the projects that place the most demands on you and your team are those where delays, cost and quality variances keep arising. These tend to be high value, longer duration and multiple stakeholder projects. As these components increase, so does the demand on the project manager. Bigger projects warrant project managers for different sections of the project, with a lead project manager directing the individual project managers.

The word 'need' has been used a lot above. As a successful project manager, it will be valuable for you to identify, work with, and capitalise on, the difference between what you *need* and what you *want*.

Putting effort into what you *want* can waste a great deal of time and energy. To optimise your time and energy focus on actions that you *need* to complete. Putting effort into what you *want* can only be considered if there is an amount of energy left that you are deliberately prepared to use on things that you *want*.

What does this mean? Well, for example, you *need* oxygen, food and water to live and to work. Depending on your projects you may *need* a basic car to get to around, and between, your project and team member locations. You might *want* a Jaguar automatic in sparkling gold, but if you cannot truly afford it, it may be that you only *need* to obtain a basic car until you are feeling wealthy (... not many Project Managers are wealthy - maybe you should try another profession or the lottery).

Your client and project *need* you to focus on effective and efficient ways forward. You *need* to innovate and you *need* to encourage carbon reducing ways of implementing the solutions relevant to the project goals. You don't *want* to spend time and, therefore, money building a new expensive concrete tank, if there's a disused one available on the site. You don't *want* to get a new costly survey carried out on the site in a hurry – you *need* to find if one already exists, maybe from a previous project. Differentiating between these two words focuses your mind, and with it, your team, on the important actions that will deliver your goals efficiently.

This chapter on goals has been positioned in the early part of this book, as experience has shown that goal setting and goal achieving are daily, hourly, and a by the minute focus for successful project delivery. This in turn is linked with good time management. You will find various references on goal setting and time management throughout your career. As attitudes and ideas develop, new books are written by different authors who explore each of these from varying angles. The ways in which we work constantly change, and with that change comes new ideas on how to manage projects and ourselves.

One methodology developed through experience is SMARTER goals – in which Goals should be:

Specific ~
Instead of 'I want to do that report today' be specific with -
'I am creating a draft report by 3pm today'.

Measurable ~
Using the same example as above you will be able to measure that you have achieved the defined action at 3pm today.

Actionable ~
Procrastination is a risk to many projects.
Identify an effective step forward and take it.
Taking action enables you to move your project forward.

Realistic ~
Setting the target too high can prove demoralising.
If this happens then temper your next action.
Be bold, but be wise.

Time limited ~
Goals with no time limit have a habit of dragging on and on...
Set out to achieve. Achieve. Move to your next goal.

Enjoyable ~
This is your job. It's what you do. Making your goals, and actions, enjoyable can make each day worthwhile, and great to be part of.

Rewarding ~
As you achieve actions - reward yourself and your team to make your working day feel positive.
A positive reward makes goals worth achieving!

The term goal has been used above to describe different day to day aspects of a project, as well as the clients overall goal of the project. You could describe the collection of tasks in a goal as an Event instead, with your Project being a collection of Events. However, popular terminology currently uses the term goal in both these contexts, and as this is an introduction to project management, it's best to adhere to standard terminology.

Summary for Goals –

Agree a definition of what the project should, and should not, deliver.

Need and *Want* – Differentiating between these two words focuses your mind, and with it, your team, on the important actions that will deliver your goals efficiently.

Your client and project *need* you to focus on effective and efficient ways forward.

Risks

As you start your project, you will almost immediately encounter risks.

What is a Risk in the project context?

A Risk is something that could deflect your project from its time and/or cost and/or quality defined goals.

To be a good project manager, you need to be prepared from the outset, in developing your project, by considering what might be a risk to it. This may sound counter intuitive – you haven't encountered any problems yet, so how could you know what might be a risk to your project. The reality is that if you don't put energy into forecasting what may trip your project up, you could be so unprepared that all you could find yourself doing is fire fighting each risk as it appears. To enable you to resolve these risks, some of your resources will be directed to working on them. Meanwhile your remaining resources will be trying to deliver your original project without good quality input from you. Your project could find itself pointing in all sorts of directions, in numerous pieces. This would need you to stop to take stock of the situation and to get the facts. From there you would need to develop an understanding of where you are, review where your destinations are in relation to your current position and plan how to close that gap. You could then need to put together a report to take to the project board to explain the situation and its effects on the project. By comparison, providing yourself and the project board with a list of potential risks, timing and cost potentials, from the start, would enable you to get them on board to the risks, ready for action to be quickly agreed on the ones that mature. After all it is their project as well as yours.

It is not being proposed that you have a thorough detailed plan ready for everything you can possibly think of that may happen. You can't. There are limitless potential risks to consider. You cannot be expected to plan for every possible one. To do so would take up your entire life and stop the project moving forward. Instead, simply consider what may happen on your project in the set of circumstances relevant to it.

The risks being suggested that you plan for could be, for example, archaeology. As project manager on a tunneling project in a historic city, a risk to consider could be of archaeological artefacts being found. Your risk management plan could comprise, in part, engaging a consulting archaeologist. Their brief could be to attend site each time the contractor was to excavate the first metre depth of each access shaft for the tunnel. Through this your client has therefore been made aware of the risk. The second part of your risk plan could be that if any artefacts are found, then you, the archaeologist, and the contractor are to discuss the way forward to create actions focussed on caring for the archaeology, while defining a way forward for the tunneling works. The contractor can, on reviewing your risk log with you, see and commit to the need for him to have a part to play. The archaeologists would be aware of the commitment you, your client and your contractor have to archaeology.

From the basic position that there was shared risk awareness, it would be practical then to escalate the archaeological effort. The contractor would be able to act to create a solution to suit the needs of the situation. That would be sufficient for the project plan. That's what you need. It's all that you need. By comparison without thought for the risks, getting to site with no prior thought of this could result in finding something, stopping work until you could get the attention of the archaeologists, arranging a team of them to act on the problem – having disrespected them by not considering their role in the first instance. The project manager would have had to notify the client that this unforeseen problem was in the way of the project, and could they afford to live with this unforeseen risk impact? It could easily have become a difficult scenario. But by considering potential risks, the ongoing mutual awareness would enable an organised, efficient and effective response.

Capture your risks in one place to keep an eye on them – your Risk Log. Be prepared for any risk to mature i.e. to escape from your log and out into the real world. Have at least one idea per risk prepared in readiness to consider applying so that you have a potential quick start for each risk should it come to reality.

As indicated earlier a way of aiding yourself here is to share awareness of the risk, and its potential for damage, with your team. They will be able to watch out for actions *they* can take to prevent it breaking loose. Someone in your team may be able to act on the escaping risk quickly, stopping it early, rather than finding that you are caught up in a fire-fighting situation. Being able to respond to maturing risks on a prepared basis is very empowering.

When defining an idea of how you could address a risk, you should also place a value on it in your risk log. This value is an assessment of the impact on your project budget and time. A summary of risks and their values, from your risk log, can be shared with your client. In turn your client can then be prepared, with a sense of what each risk could mean to their wider business, so that they can be ready with informed responses. Your client may want to address the risks head on, rather than wait to see if they occur e.g. they may reshape the goals in some way to reduce the likelihood of those risks. They may take a risk out of the project by dealing with it up front in another project. If you don't inform them your client cannot have that opportunity.

You may find it practical to categorise your risks. Doing so can assist you in managing them. Category examples for your project may be –

- Health and Safety

- Environment

- Financial

- Quality

- Contractor

- Client

- Consultant

- Suppliers

- Designer, etc to suit your project.

Incidentally, it can be very frustrating when someone asks 'Do you have a risk log?' They need to clarify their question. Which of the above risk types are they asking about? Clear communication is very important.

No matter what category your identified risks belong in, the great positive aspect is that you are aware of the potential for risks to occur. Having awareness of that potential enables yourself, and others, to be prepared to take a planned course of action. Although others may not want to be prepared - some people act as though by being kept in the dark, then they can blame somebody else. This is the opposite of good teamwork and is not good business sense. It contributes to lack of clarity in what to do to prevent the risk maturing, as well as in what to do if it does mature. Be aware of such people as their limited competency is likely to lead to them blaming everyone except themselves. Such people may go on to cloud the situation and its solution – making the problem even more difficult to resolve.

It is said that information is power. This is particularly so in respect of risks. Use your own awareness – log that risk; assesses it; monitor it.

Risks may decrease, or increase, in terms of their threat to the planned successful project delivery. Their scale, potential business impact, possible cost and likelihood will most likely change during the life of your project. Monitoring them on a weekly or monthly basis will help you, and the wider stakeholders, to be mindful of them and of the damage they can do should they escape from your risk log into the real world.

Despite all of your preparation for risks, when they escape, or one arrives that you had not pre-considered, handling them can still be stressful. The more risks you have the more that stress can build up. Conversely, the more you share awareness of risks, the more prepared you are. Those people you shared them with can't shout that they didn't know. This relay can serve to cut your stress levels and is, therefore, highly recommended!

Sharing awareness of risks can be amazingly productive. Ideas, support, and preventative actions, can surface from all sorts of directions, some of which can be entirely unexpected. Such support ranges from advice and guidance on how to manage / remove/ control/ value the risks, to actually taking the risks off your project. Remember that old adage 'a problem shared is a problem halved' - this can definitely apply to risk management. Ask yourself the question 'Do I want all the stress of unforeseen risks?' If you do then go ahead, knock yourself out. However, if you want to share the impact they have, you can choose to work in ways that get that target off your back...

Summary for risks –

Log each one.

Consider how to deal with each risk should it mature/escape.

Share knowledge of the risks.

Log that risk. Assess it. Monitor it.

Issues

Issues are, typically, questions asked about your projects that need answers.

You will benefit from having a place to record and monitor them all – your Issue Log.

Incoming issues are like darts thrown at a dart board. Without an Issue Log, *you* become the *dart board*. With the Issue Log in place, the darts land in the Issue Log. You look at each incoming dart and obtain facts pertinent to it in relation to your project. Calmly hand the dart back to the thrower, providing them with the new information attached, and thanking them for their enquiry.

When an issue gets resolved within the project time, cost, quality boundaries then you can close it down. If it doesn't fit inside those boundaries then you will need to determine whether your project should, and can, handle it. Can it be closed down as being not relevant? It may be appropriate for you to decide to regard some incoming issues as new project risks. You can move those to your risk log and manage them as risks from then on.

While each one is being progressed towards resolution it can be copied to your ToDo List, or preferably someone else's!

Most issues can be considered, answered and, therefore, closed within your project boundaries. Your Issue Log should note who raised it, what it was and what happened to it. The issue log enables you to record the relevant answer, and that the answer has been fed back to the originator and has been closed. Keeping the answer permanently in the log can pay dividends later. It's surprising how many times a question, or similar one, gets raised. A quick skim through your Issue Log will remind you what the answer was. You send the same answer to the next enquirer. Job done (again).

Without the Log you can find yourself, and/ or others in your team revisiting a question to rediscover the forgotten answer – or even worse to find a different answer which causes rework problems of revisiting the question again etc. We are all busy people. Remembering the answer can command a lot of your memory, time and effort. Reviewing a log to re-find the answer is less stressful, less time consuming, and enables you to spend valuable time, and therefore money, on the important tasks (you'll have plenty of those, so you could do with a self help tool such as an Issue Log).

Summary for Issues –

Log each one, permanently.

Review the log to check if you already have the answer.

If it can't be closed down within the project boundaries' should you transfer it to the risk log?

Mind the Gap

Despite all your hard work, and attention to detail, something in your project will almost certainly cause it to deviate from your project plans. When it does, you will need to determine what actions to take to get the project back on the rails. These actions can send you off in all kinds of directions using up time and money. As though you were not busy enough already! You could try to go back to an earlier position in your project to find a way of getting the project back to where it was before the situation occurred. Only to find that moving forward again takes you, via a different route, to the same dead end. While you were going back on that task what was happening to all the other project activities. You need to keep working on those too! How do you sort this out?

One highly effective, and efficient, strategy for getting through that situation is –

Hit the _pause_ button!

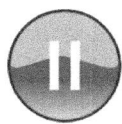

Despite the pressure to dive back in unprepared – look *forward* to confirm where your needed destination is. Assess the Gap between where you have paused, and your destination.

Define the effective actions needed to close that Gap. Then deal with that Gap.

Play …

This simple process can be summarised with the term 'Mind the Gap'.

Burning time, energy and money going back, and attempting to come forward onto your original preferred route can keep you busy for hours, even days or weeks! Meanwhile something else will be likely to be going off balance. Take action only to close that gap. Keep moving forward.

> *Summary for Mind the Gap –*
>
> Assess the situation; where are things?
>
> Where do they need to be?
>
> Move forward not backward.

External Processes

In order to deliver your project you are likely to need to work with, through, or around, a number of people within other businesses or organisations beyond your own.

These other people, and other businesses, will have their own business or organisational goals to meet. They will have their own ways of working. To get their business goals met, these bodies will need to fulfil business or legal processes in delivering the needs of their businesses. Some of their ways of working, or their timings, are likely to not match your project requirements. As a result some of these may negatively impact on your project.

In planning your project you could consider, and make assessments within your project for, processes within those external businesses. To some extent you are definitely planning for the unknown here. It's like shining a torch into an abyss. The good news is – it's you that's shining that torch.

You will need to find out what these other parties _need_ to do, or _are going_ to do, which will influence the successful delivery of _your_ project. Look to understand the impact of those actions, or inactions, on your project. Identify solutions, or workarounds, that will move your project forward. These may present themselves as new requirements adding to your project time, cost or quality. Maybe you could capture some aspects of these processes within your risk log, to enable you to monitor them. Would it be relevant for you to alert your client to identify how they would prefer to address such risks? Consider what your team can do to contain their effects within your project parameters. Look for the minimum input you and your team need in order to keep your project moving forward towards successful delivery. Plan this input, and get relevant parts of your team that can contribute to the solution, aware and actively contributing.

Putting effort into redefining your project in some way may be necessary.
At the end of the section on risks it was identified that you could define a risk for unknowns. External business processes are unknowns that, to only some degree, you may be able to quantify at the outset of your project. Potentially they could be in your risk log to the extent that in fact you would have no known idea for dealing with them...an unknown!

Summary for External Processes –

You are likely to need to work with, through, or around, a number of people within other businesses and or organisations beyond your own.

Find out what these other parties *need* to do, or *are going* to do, which will influence the successful delivery of your project.

Identify solutions, or workarounds, that will move your project forward.

Plan

Why do you need a plan?

As the project scope increases, in the absence of a plan, you will need to rely, to increasing degrees, on your memory. A memory that is busy with many tasks and details. Anyway you are going into a meeting and don't have time for this new problem right now... The bigger the project, and the more projects you are managing, the busier the inside of your head is going to become. Without a plan it will be impractical to engage others in your project needs. It will be impractical to keep yourself focussed. (Was that a headless chicken that just went by...?)

Chaos is easily achieved. Organisation takes time and effort, but gives visibility. Visibility provides the opportunity for control, or at least as near to control as you can get. This visibility can be shared with others – within your own team and with the wider involved businesses /organisations/ bodies that you are likely to encounter.

Your plan can vary in detail from a few bullet points on a page, to a full critical path plan such as what you could produce within a formal planning tool. How extensive your plan will need to be will be influenced by the scope of the project and the number of other people, businesses, organisations or bodies, who you determine would benefit from either seeing your plan or contributing to it.

Some of the best plans are the simple ones. Yet there is no doubt that larger projects result in extensive, intricate, plans. Whatever your plan scale it must, in the end measure, be effective in supporting the achievement of your project goals, with the minimum of input. If you spend days creating and delivering the plan, then it's likely that you have forgotten to focus on getting the job done. The more intricate and detailed you make the plan, the more time you will need to spend creating, and updating, it rather than delivering it. Yet failing to plan is undoubtedly planning to fail. Keep it simple, so it can be easily referred to, and updated. A simple, clear, easily updatable plan is a powerful tool. Include the project objectives, to link everything to delivery of them.

You can only plan what you know today. Your plan is a flexible adjustable device that you can add to, and guide yourself from, as your project progresses. It is not normally a document that is created and never updated. It is normally a device that you will add detail to and amend as more information becomes available.

Once you have a plan you can test out impacts of incoming changes, or proactively test what may be the impact of potential risks. You can share sections of your plan with colleagues to determine if your proposals are realistic. You can develop the plan, with the input of others, to improve the detail needed to achieve a good degree of confidence in delivering to your timeline. By developing your plan you can see where more resources may be needed to achieve the timeline you have available. You can allocate costs to each of the people and actions, enabling you to plan expenditure. If you plan the spending in your project then you can develop a financial budget. Through this visibility of a financial profile you can demonstrate the needed funding to your client. This will enable your client to plan to make the needed funds available at the right time to deliver the project. Then you and your client can pay the projects bills in a timely fashion. Paying bills will keep the various bodies in the project committed to your project. You and your client can determine if the project has the funding it needs overall. Your client can be clear about the commercial viability – or otherwise – of the project commitments. It's always possible that they may cancel the project if the numbers don't give viable financial returns. If that's the right decision then at least it can be made, rather than finding out when the project is part done and there is no money left. Knowledge available through developing a plan enables you, your client, and involved bodies, to be informed. With information comes power. The power to manage your project.

When your plan isn't working, be honest with yourself; update the plan to make it realistic. Then keep moving forward. Many plans are noticeable for what is *not* on them. It's often what's *not* on your plan that will bring your project towards failure. When something appears on your radar that becomes apparent to be needed in your project, but isn't in your plan, add it in to see how you can close the gap – your plan will be more realistic for it. It's the realistic plan that is the successful one. But be cautious - you can flood yourself with the nth degree of planning and not get anything delivered - do only enough. You don't need a tome. You *will* benefit from a realistic plan.

One of the great realities of project managing is that there's likely to be a 'jack in the box' out there somewhere (or a whole load of them). How do you plan for a burst water main in your site? One approach that does help here is to ask relevant potentially involved people, internal or external, for input and comment, information, ideas, expertise. It's often amazing how many ideas come to the surface about how to work around challenging problems. As project manager, you need to encourage the right people to get together and propose solutions which you could support within the time, cost and quality boundaries ranges of your project. Put your effort into keeping the client informed of this new 'jack in the box' and its potential impacts. That's the way it needs to go sometimes. You're not omnipotent – you're only the project manager – you need to lead the right people towards finding a viable solution that the project needs.

The best delivered projects are the best planned ones with the best motivated, talented, people on them.

Summary for Plan –

Keep it simple, so it can be used easily and can be updated easily.

Include the project objectives, to link everything to delivery of them.

Your plan is a flexible adjustable device.

Budgets

Your client has a defined budget for delivery of their project through you and the wider team. They are paying the bills. Their budget will be related to the worth of the delivered solution in its finished, operating state. They need to get the work done to the defined price – or they may not be able to secure their return on that investment. On your part ignoring that budget will not make it go away. You are tasked to enable the goal of delivering the project outputs, one of which will be out-turning cost stability. That's why you will need an overall cost/ budget, comprising a number of sectional budgets. Yet an accurate, fully costed, individual budget for any given activity can be difficult to secure. While some can be quite clear. As a result an accurate overall project budget can be very difficult to both establish and to maintain.

Your project risks will affect your project budget. This compounds the difficulty of establishing a firm overall project budget. The overall budget needs to contain a realistic list of items required and their costs to deliver. It's easy to create an incomplete list, or to undervalue the items on the list. If you think it may be a good strategy to over-forecast values, then the budget could over-run the clients available funding – and that could, unjustifiably, stop the project proceeding. Sounds easy – so what's the difficulty?

The difficulty tends to lie in the bits that are missing. And in the parts that will not go to plan – the risks.

Include in your budgets for everything relevant that you determine you will need – people, materials, delivery, testing, investigations, reports, business processes, changes or corrections, approvals, research, development, trials – well, everything the project will need to take it to its goals. The more realistic and balanced your costs and risks build up, the more likely you will be to deliver to budget. Look for sources of information, and make judgments regarding scope included and items excluded.

Building up a cost for materials and staff is difficult. Material prices change due to all sorts of local, national and international factors. It's likely that none of these will be under your direct control. So how can you be expected to be in control of your project costs?

Staff costs vary with the individual and their circumstances. Consider how long each person will need to deliver their part in each task. Allow for inefficiencies, human error, individual and collective learning. Refer to others for their experience of how long a task may take. Assess their confidence in their answer. If your project spans a salary review date, plan for a potential salary increase.

It may be appropriate to include a contingency – a budget for unknowns. None of us knows everything about everything. Planning something for unknowns is, simply, realistic. Your client may choose not to budget for these. Giving them the opportunity to make that decision gives them the power to deliver the project from an informed position.

It's very likely that one or more of the individual item costs you have set in the overall budget build up, will over-spend. It is also possible that some will under-spend. Your challenge is to keep within, or up to, the overall agreed and funded budget. Look for savings in all of your planned items and tasks. Challenge the individual budgets – don't just accept them. Allocate parts of the budget to those who are actually spending it – team members, contractors and suppliers.

The more sections of the team that can take ownership of their share of the overall budget, and deliver to it, the more likely you are to get your project delivered inside its overall budget. This can take the pressure off some over-running costs and so contribute to keeping the overall budget in control. When your efforts to maintain balance in the overall budget show that the forecast cost is escalating, then it's time to inform your client. They need to know, to enable them to decide the way to direct the project. Show your client the current out-turn forecast for each item, against the original proposals. Know why there is a difference. Update your risk log and determine the remaining risk cost on the project. Understand how this affects your forecast out-turn cost. Together these will give a good awareness of the overall budget.

Transparent costings enable you to have good clear manageable control of the project, and of where it is heading. Maintaining control over the financial status of your project gets you, and keeps you, in a position to be ready for unexpected events. When a new problem arises – quickly share the fact that it has, and what its impacts are. Be ready to inform your client of the project status.

Things will not always be what you plan to them to be. Something out there will not arrive in budget.

Something out there will not land in time. Something out there will not be in quality. When you see it coming towards you, figure out how the overall project can still be met, and work to resolve it. If you cannot get the project forecast to stay in the budget then work this fact through with the client and the wider team. Find out what must change and get that change driven into place! Keep closing that gap.

Summary for Budgets –

The overall budget needs to contain a realistic list of items required and their costs to deliver.

The more realistic, and balanced, your costs and risks build up, the more likely you are to deliver to budget.

Challenge the individual budgets – don't just accept them.

Control

If only!

Depending on the size and duration of your project, you may encounter –

- large numbers of people inside your organisation and external to your organisation

- numerous businesses, utilities, councils, service providers

- many suppliers

- multiple clients

- members of the public, environmental bodies

- local, national and international issues

- risks

When you add this up, the likelihood of you being in control becomes, in some ways, quite small. Yet without your leadership there can only be chaos. You have the role of project manager. In this your team members look to you for leadership. For you to be the leader they need, you will need to apply some form of authority and control. In reality what you actually need is for others to be informed and for you to have an effective system to operate within. Your plan, risks log, issues log, time, cost and quality goals and budgets give you this.

You cannot control what other people do, or the organisations in which they work. Knowing this, in itself, gives you an awareness, or knowledge. As mentioned elsewhere in this book, knowledge is power. What power in respect of this section of the book? It is the power of knowing that you have only *some* control. It's how you apply that amount of control that gives you the focusing leadership position in your project.

What can you actually control? Often the answer is only *what you personally do*. When managing multiple tasks across multiple projects in order to drive events forward, you find yourself relying on the actions of other people, plant, materials, financial activities and processes. Yet because most people you encounter are good, willing, team players, what seems like the impossible can often with a little leverage, be achieved. Liaise with others. Involve others in your plan. Commit to the part you can play in the plans of others (getting something achieved is often a two way activity).

Throughout all of this be aware that the more you look to apply rigid control of exact activities, the less you could find yourself being in control. In fact much of what you can do is to *influence people and events*, rather than control them. This is because other people have their many tasks, on many projects, to implement. It's also because everybody is different. Their difference needs them to have direction, in order that they can apply themselves – their way – to actioning the goal you have delegated to them.

An intricate web of activities is in existence every day. The more you attempt to control each interface of this web the more you could be stretched, and the more your projects and you could become at risk. Contrarily, the more you seek to apply the realities of this big picture, glued together with an amount of flexibility in how you inspire your team members, the more that can be achieved.

Control, and the application of it, can attract stress. The more control you attempt to apply, and the less of an integrated system in which you seek to apply it, the greater the stress you can find yourself carrying. Your people can help. The more you know the people with whom you work, and the closer you can work with them, the more they find the capacity to deliver what *your* project needs.

Carrying multiple projects requires, generally, progress to be achieved through many people across a number of businesses. Control tends to dwindle in these circumstances. However, the structure of shared goals (time, cost and quality) helps everyone to focus their input and to commit themselves in, at least, one direction. To maximise your project progress, you need less control and more positive enabling people skills. Skills focused on motivating, listening, supporting, encouraging, training and coaching.

Summary for Control –

For you to be in Control you need others to be informed.

Much of what you can do is to influence, rather than Control.

Shared goals (time, cost and quality) help everyone to focus.

Carbon in the project Environment

Your project makes a difference to your planet. It's your planet. What are you doing about it?

Since the onset of the industrial revolution, mankind has been impacting on its environment. Engineers, scientists and innovators have turned skeletons of an idea into life changing creations. This has come at the cost of local, national and global changes to our environment.

The Intergovernmental Panel on Climate Change was established in 1998 to report on increasing scientific concern over the possibility that humans are causing significant climate change. Their most recent report (in 2014) states –

> 'Human influence on the climate system is clear... human influence was the dominant cause of global warming between 1951 - 2010.
>
> Atmospheric concentrations of carbon dioxide ... have increased to levels unprecedented in at least the last 800,000 years.
>
> The overall risks of climate change impacts can be reduced by limiting the rate and magnitude of climate change.' [5].

Through our determined growth and development humankind has been placing demands on our ecosystem that are outstripping its ability to keep pace with us. That's bad for your planet, for you and ultimat-

-ely for your client. The immediate guidance on this is to cut carbon emissions. Hence the drive to reduce carbon production in the life cycle of your project. How can you cut carbon emissions while delivering your project within budget?

There is carbon embedded in the production, delivery, operation, decommissioning and disposal of project items. In many projects the needed investment involves excavation of raw materials, manufacture of plant and components, shipping, construction of structures and maintenance. The information from global representative bodies is that those capable and able of imagining and actioning carbon reduction should lead the way in reducing carbon as an active aspect of the delivery of their projects, to address the needs of our planet. All of this can impact on our planets capacity for self maintenance and the longevity of its future.

Clients are, as a result, becoming increasingly active towards driving down, and ideally driving out, carbon from the design, construction, operation and maintenance processes required in bringing their project to fruition. How do you measure this? What should be measured? What budget should be included for it?

Knowledge is currently developing about how to do this, while action is also being taken in parallel with this development of knowledge. A range of support is being developed on the subject of measuring and reducing carbon in project lifecycles. There are indications that well managed carbon reduction effort in a project, can save cost in the project lifecycle. There is, currently, no single means of measuring carbon in respect of any given project activity.

You will need to look to peers and the market to guide you on this journey – your planet is depending on you and your clients' attitude towards carbon reduction for its future.

In making carbon cost decisions, information is developing for you to equip your team to look at both carbon reduction and the financial cost of carbon reduction in selecting solutions to designs, construction, operation and maintenance. Saving carbon and cost can be made to be a double win for your client and your project. Some capital intensive companies are developing in-house carbon reduction and interconnected capital cost saving models. Your project can gain from application of such a tool if it is available in your business.

Achieving your project goals, while sustaining your planet, can make sense for shareholders and the planet alike. A healthy planet will be able to provide a healthy future. A healthy future should provide a wealthy future.

Summary for Carbon in the project Environment –

Your project activities make a difference to your planet.

Carbon reduction is proving to be an important aspect of how to address the longevity of our planets future.

Look at the financial cost of carbon reduction.

Chasing

In an ideal world we would have all our work to do and it would be precisely time planned and carried out to the second. In the design, procurement, construction and commissioning effort of a new project, this can prove to be an immense challenge. The reality requires that a number of activities, for various reasons, need to be re-planned. These re-planned activities cut across the arrangements of other activities. With this going on over numerous people in many businesses, the impact of these changes needs to be communicated to the project manager. Formal reporting should make this visible, but there is nothing more effective than pro-actively chasing to find out what is going on. If you don't chase up and monitor, then you would not be aware of potential slippages, until it's too late for you to assist with a recovery plan. Once it's too late the overall project time, costs and possibly quality content could be negatively impacted. That's bad for your budget, your client and your project.

So as a project manager you need to chase. Not yet – I haven't finished typing...

Chasing needs to identify what is at risk of slipping. Your objective being to assist in removing whatever is causing the slippage. If you don't do this, your project and, therefore, you, will in some way and to some extent, fail. Chasing doesn't need to be aggressive. It doesn't need to be disrespectful. Above all it doesn't need to be abusive. It needs you to recognise that each busy person has much to do. Some of which may not progress according to the proposed plan in respect of your project.

You need to get your people focussed on what you need them to do. You need to confirm they are doing what they have said they would do. Other project managers will be checking with that person in respect of the work they need done for other projects. Your project is in competition with others for their attention. This places a risk on tasks and events you are expecting to have done.

Chase the actions you need done. Assist them to get them to move forward. Get them closed off. Tasks do slip – due to many things – some of which reflect the depth of detail in the task. Once you find out about the slippage you will be in a position to re-plan, so staying in control, rather than getting bad news when it's too late. How you handle this bad news with the person involved could influence how they prioritise your tasks in the future. If you bully that person you could find that the results you receive from them in the future are consistently not what you expect. If the person is struggling, can you get them some help, or some relevant training?

Most people are fundamentally genuine and want to do a good job. But such people tend to get put upon until they buckle. At that time they deserve to be supported to get them back up to their capable productive standard. After all, they can gain from finishing what they have to do for you as doing so will leave them with one less job to do for a while, until they get their next new task. Positive minded people should receive skilful chasing as a prompt to help focus them to clear their task.

Everybody is different. When chasing, some people may respond better to emails, to phone calls, telephone messages or to face to face contact, or texts. Some people need combinations of these. You need to explore what works for each person.

With large amounts of information, sending an email may not be enough. Booking a short meeting with them to explain the detail to them may help. It's up to you to try different approaches until you find the one that works. Some people may tell you how it works for them – if they know. Chase it. Assist it. Close it.

Summary for Chasing –

Your objective being to assist in removing whatever is causing the slippage.

You need to get them focussed on what you need them to do.

Chase it. Assist it. Close it.

Keep moving forward

Your workload, and some individual tasks, can sometimes seem insurmountable. Planned, and unplanned, tasks can sometimes accumulate into a seemingly endless list of activities, implications and risks. It can sometimes be unclear what task to start with, or where or when to start. The daily variety of project communication, information and data can create an absence of clarity of what to do next.

With all the actions that you are juggling, how do you capture and address today's new tasks?

Some form of personal process that will get your actions logged, monitorable and actionable in an effective sequence is called for. A simple ToDo list, through to any number of a variety of systems and methods are available. Such a ToDo list should be relatively quick to use, to save you losing track of the action! Multi platform ToDo list apps such as Wunderlist may be useful here (**www.wunderlist.com**). Keeping a notepad at hand for quick notes that you can follow up on can be very useful. Evernote is a multi platform app enabling you to capture a note on one device while picking it up later on your PC to incorporate, or adapt, into a word document report (**www.evernote.com**).

Time management training can help. Usually books on time management advise about giving tasks a priority. However, there are times when pretty much everything has the same priority. Then what? Ask yourself some questions – try starting with –

- Does it need doing now?

- Who could do it now?

- What resources are needed to fix it?

- What ideas do others have?

- Where does it need to be done?

- If it was left till later what would the time, cost or quality implications be?

- Could it be managed as a risk or an issue?

- Can it be done in parallel with other tasks?

- Does it need to be done following other tasks?

- Why does it need to be done now —when else could we do it?

If it can be done now, to the benefit of the project or events or tasks – get it done. If others can do it now without significant delay – get them to do it now. If leaving it till later will cause significant knock on delays or significant cost increase then do it now.
If it's a number one priority then it must, logically, be right to action it. So do it.

When it is unclear where to start, perceive the looming mountain in front of you as just a collection of rocks. Deal with one of the priority one rocks. As soon as you remove one of those rocks the size of the mountain gets smaller. Several rocks later and there's only a hill where the mountain used to be. A few actions, or rocks, later, there's no mountain and no hill any more. Enjoy the feel good factor as you break your mountain down!

Reward yourself, even if it's just a tick on your ToDo list, to show yourself you have been making a positive difference today.

Keep achieving. Being positive helps many things get resolved. The more projects you juggle simultaneously, the more failing tasks you are likely to encounter. As you carry more and more projects simultaneously, a time could be reached when the majority of your energy is spent examining and contributing to the fixing of failures (the things that *are* working – through your efforts - tend *not* to need your attention). When fixing things that are not working dominates your time, you need to keep yourself driven.

You need to be your best friend, and keep yourself smiling. After all, if you were not being strong, identifying and driving the fixes forward, then the fixes would not happen and your project goals could not be achieved. Despite at times only dealing with problems galore, the role you're implementing is valuable to your project to enable it to succeed. You are acting to bring about change. Sometimes it's easy. Sometimes it even happens, through others, without your detailed input. When it's an uphill struggle, focus on your goals. Enjoy the success of achieving each task. Celebrate achieving your project tasks and goals. Overall – enjoy every improvement that you make – you deserve to, because your project would not have got done without you. Be proud, be strong and be positive.

Summary for Keep moving forward –

It is a number one priority and it must logically, therefore, be right to action it. So just do it.

Enjoy the feel good factor as you break your mountain down!

Be proud, be strong and be positive.

Time, Cost, Quality

When delivering your project, each task that needs your attention may involve the juggling of any or the entire three project boundaries set by time, cost and quality. Looking at each task with the aim of fitting within these three boundaries supports you and your team members to keep your solutions within your project goals, and keeps you focussed on your clients' needs.

If your solution, in balancing these parameters, means there is a need to exceed one of these boundaries then you have a sound basis to raise the solution, and its project impacts, to the client or project board for approval. Reporting to the project board with a solution founded on these three parameters will validate why your recommendation was being made.

With 'sleeping' clients (those who prefer to not get involved) balancing these parameters can be really tough. Get ready for a rollercoaster ride. Be clear about the facts (don't use guesswork). Be clear about the ways forward (don't create a fog for yourself to wade into – it's cold and clammy in there).
As your project progresses your team should become aware of relevant priorities around these three driving boundaries, and should develop the skills to create solutions to problems that contribute to these. Should events occur when the known scope for the three parameters will not be sufficient to drive an obvious decision then look at the situation as it is evolves. Use the situation to extend definitions and then make recommendations around that new data. Consider cha-

-nging the lengths of each side of the triangle to find your solution. Include your client in the problem of the time, cost quality balance, to identify their views on your identified ways forward.

The nature of a project is such that aspects about it will not fit with what you can read in textbooks. People around you can be a great source of ideas, knowledge or inspiration. Ask how your team members and colleagues can help. What experience can they add to the situation? Even negative experience can be helpful as a guide of where not to go next.

The length of each side of the conceptual triangle is rarely cast in stone. You could find yourself working hard to keep the lengths of the sides of your triangle firm, but when risks mature, tough decisions need to be made. Creatively consider all the information available, and define the information not available. Compile the needed actions to enable your project to keep moving forward. Then develop your recommendation.

Summary for Time, Cost, Quality –

Looking at the task with the aim of fitting within these three boundaries keeps your solutions within your project goals, and keeps you focussed on your clients' needs.

Look at the situation as its evolving.

Change the lengths of each side of the triangle to find your solution.

Materials

Many projects have an inherent demand for various materials. As a project manager you will become involved, when establishing your budgets, in the cost of materials. Will the price of materials be stable during you project lifestyle or will they move, causing impact on your budgets?

In recent years there has been press coverage regarding the availability of steel. Sounds like a dull topic! Yet project managers delivering projects involving steel suddenly found themselves at risk of not having sufficient steel available for their projects in time. They found themselves with a _risk_ to the delivery _time_ for their projects. With limited availability came a risk of increase in _cost_ of the steel. Could alternative designs or materials be used to get back on time or in budget? This in turn would require a change to _quality_ for their project. In just this one material there was potential to impact time, cost and quality.

Why was this happening? The growth of China, and the effect of its economy growing so fast, was such that their market had a demand for steel that caused global steel markets to begin to be impacted. Market prices for steel increased beyond any forecast that project managers around the world would have planned. Supply of some steel products became delayed. Project managers found themselves looking to review budgets, and to possibly re-plan some aspects of their projects.

Global activity may seem unimportant to you on your one project. Yet the preceding example indicates the effect that wider global events do sometimes have on UK projects. When the cost of oil increases so do products that involve it in their production cycle. What could you do to mitigate such risks? At the outset of your project might it be appropriate for you to include a risk in your log that considers market demand impacts on materials?

Dependency on a narrow range of materials can leave you with an 'all your eggs in one basket' scenario. During development of your project could you encourage the team to consider a range of alternative materials in the project to spread the risk of materials demand? Could the push for carbon reduction take us away from this or towards it with many rushing for an alternative which then in turn becomes in short supply?

When foot and mouth broke out in the U.K. access to some areas became restricted. This prevented the movement onto some sites of some materials where the site was too close to affected farms. Delivery of materials was at risk of delay. This really is such a rare risk that it's one of those not realistic to normally plan for. As stated in the chapter on risks – you cannot be expected to plan for every possible risk.

The goal in introducing this subject of materials to you in its own chapter is to bring your attention to the reality that materials, although a seemingly unimportant topic, can have a significant impact on your project. They impact on carbon emissions, time, cost, quality, risk and planning of your project.

Summary for materials -

Materials, although a seemingly unimportant topic, can have a significant impact on your project.

In just one material there was potential to impact time, cost and quality.

Might it be appropriate for you to include a risk in your log that considers market demand impacts on materials?

Teams

As described earlier, you are likely to encounter a number of bodies such as contractors, consultants and suppliers in the delivery of your project. Can you get them to be part of your team, to function efficiently in your project? They will have their own business objectives, just as your employer will have theirs, and your clients' business will have theirs in turn.

The balance of business risk between each of these bodies will influence how each body behaves in response to each project risk and issue. Contractual arrangements that place all of the risks with your client tend to bring about solutions which result in the duration and cost escalating on your project, as the matter of solving problems then always rests with your client. Instead, distributing risk in a balanced way across the contractual bodies can bring about dynamic, challenging, productive behaviors' from each body. In turn, making these bodies part of your team brings in their ideas to the advantage of the project. The goal here is to make the balance of risk move to the team members nearer to the risks. This brings solutions that challenge the team to keep project time, cost and quality in check.

Making as many of these people as practical feel a part of your team can help you deliver your project. You can drive this by involving external suppliers in the timings of your needs, or of actions you need them to complete as part of your project. Through this contact you can identify their needs and their difficulties. Potentially you may be able to re-organise when you need outputs from one or other of your suppliers to assist them deli--ver for your project. This visibility of your supplier's capacities and timings can inform you, to enable you to improve the accuracy and robustness of your plan.

Encouraging everyone in the project to keep you informed of issues or risks keeps you in some control. Once you know that part of your wider team cannot deliver to your needs then they and you can agree what, when and how they can deliver. In turn you can then re-plan your project, and your wider team, to a new detail which enables you to get back into a new amount of control. Good teamwork!

As project manager your challenge is to make the relationships between each party, and the balance of risk position, work to the advantage of each party and therefore to the project. Understanding the conditions of engagement of each body will support you to make the interrelationships work to that advantage. Ultimately, they are in your team, and you are in control. The extent to which you can get them to be an active contributing part of your team, rather than a remote uninformed and non-contributing body, can be influenced by you.

The disparate parts of a team may have some knowledge that, if shared, could be brought to the advantage of the project. Bringing widely dispersed team members together, can pay dividends. As I mentioned elsewhere in this book, knowledge is power. In this case harnessing the power of disparate knowledge, to the advantage of the project, is the gain of enabling widely dispersed bodies to be active team members for your project.

Innovation occurs in teams. Through bringing each person to being in your team, you can encourage and listen to ideas. Innovation tends not to happen in a linear way. There will be ideas that fall flat, while others don't sound up to much, yet if encouraged may solve underlying problems or could prevent new problems coming into existence. The management of innovation can be complex. Without direct funding for innovation you may feel it's not within your remit to drive it as a goal or task. Yet customers currently insist on innovation. The clients aim being to deliver the project for less time or cost (or both). Yet innovation is risky – it's about something not done before – while your client is likely to be pressing for delivery to be right first time. It's a conundrum. Your team can innovate if you provide involvement as one team, encouragement, leadership and positive feedback.

Having worked with, through, and around others it is clear that, on a project, there are likely to be 'others'. The energy needed to address their behaviours and resulting actions can be immense. However, the return obtained in productivity and positive outcomes of working with those who are enthusiastic and competent at team working, is that your project can be turned into one that achieves, rather than fails.

Summary for Teams –

Making as many of people as practical feel a part of your team can help you deliver your project.

The balance of risk between each body will influence how each body behaves.

Bringing widely dispersed team members together, can pay dividends.

What about me

Who is the most important person in your life?

As we are all different, then the answer to the preceding question can be quite varied.

In order to deliver great projects successfully, again and again, you need to power yourself. To put yourself in the position in your life to deliver, you need to be able to focus yourself, and to motivate yourself.

When there is so much to do, it's easy to become drawn into your business world needs. In so doing leaving behind your personal, and possibly family, needs. What to do next can become very demanding and personally limiting. However, if you do not look after yourself then you cannot look after the needs of your projects, your team members or your family.

To enable you to achieve all of this, you need to be aware that the most important person in your life is ... *you.*

You are important to yourself and to everyone around you. Your health is vital for you. Look after yourself, and you can achieve all that you need to. Disregard and disrespect yourself, and you will not be able to sustain your drive, so, in turn, you will not be able to achieve your goals.

Who you are and what you can do to energise and enthuse yourself will inevitably vary. This is because the sum of what defines *you* is unique to *you*. There are numerous self-help books available that may be a source of inspiration on this topic. Your local college may be a destination you could look to for personal skills and development. There are many sporting facilities as a means of keeping fit – or releasing some of that frustration!

Web sites such as the Fish Philosophy [6], and Who Moved my Cheese [7] give easy light hearted experiences of others which may give an insight into how you and others can self manage their drive. De-stressing exercises can be useful. Once learned, you take them with you internally – for free! - applying them wherever and whenever. Great to have a copy of The Little Book of Calm available now and then too! [8].

Why mention these? It's because happiness contributes to healthiness and success at work. We have to be happy, healthy and here to realise efficiency gains at work. Happiness promotes success, according to research published in December 2005 [9]. Happy people are more likely to 'exhibit ... the attributes, resources and skills that help people thrive and succeed. Happiness reinforces positive emotion and leads to success in work, relationships and health', indicates the report. These findings suggest that happiness is not a 'feel good' luxury, but is essential to people's wellbeing. The link between happiness and success was investigated by a team from the University of California, Riverside, led by Professor Sonja Lyubomirsky.

Therefore, focusing on your job to serve both the project need and your individuality at the same time in a happy way could benefit your projects and you, and, therefore, your client.

Each person's first focus should be – themselves. We need to do something we were not taught about at school. Yes we get the Maths, English etc. We get developed to be good for business. What we don't get are tools and skills to care for the one person that can make all this difference - yourself. Yet if each of us is not happy, inspired and imbued with enough self drive to fit our circumstance, we can come to a tumbling halt.

Look after yourself.

Summary for What about me -

The most important person in your life is ...YOU.

We have to be happy, healthy and here to realise efficiency gains at work.

Focus on your job to serve both the business need and your individuality at the same time.

The future

Nobody knows what our futures will bring. One constant will happen though. Change.

To equip you for change, your experiences can add to your personal value and abilities. Membership of a relevant respected body can support you to learn and grow. As you develop your skills you may want to contribute to the growth of others. Because . . . today's learners will be tomorrow's leaders.

Therefore -

If a person wants to know   – tell them.
If a person wants to understand   – show them.
If a person wants to find out   – train them.

People working with you, who want to grow, are worth your time. They look to you for learning, knowledge and the opportunity to self-improve. Starving your team members of such opportunity stifles your business of the opportunity for it to grow, and restricts your projects of the opportunity to succeed.

Sometimes it can feel that there just isn't time to develop those around you. How did you learn? Did you get support and the opportunity to learn? Was all the support there that you would have valued?

It almost always works out that the time, effort and commitment that you put into this aspect of your work pays back - sooner or later. Often many times over.

While you are self driven and successful and supporting of others you will find that you acquire supporters around you. We all benefit from such supporters. They are the people that contribute to the success of your projects.

Everybody is somebody!

Summary for The future -

Today's learners will be tomorrow's leaders.

People working with you, who want to grow, are worth your time.

The time, effort and commitment that you put in to this aspect of your work pays back - sooner or later. Often many times over!

Health and safety

Health and Safety in the subject of Project Management is ever present. The Health and Safety at Work Act lays down requirements for employers and employees. In summary every worker has a duty to take care of their own health and safety and that of others who may be affected by their actions at work.

It's taken many years to get as good as we have got in the field of Health and Safety at work. Getting this far has cost many lives in some industries. Learning has been hard won. Yet it won't take much to undo it all, if we don't bother. The Health and Safety at Work Act has driven much valuable change into place. Life changing change! With the evolution of risk assessments and method statements, and safety being measured and driven by key performance indicators, the ultimate appalling risk of death can be, and is being, reduced.

From a cold hearted management perspective, _not_ having to deal with accidents and the fallout from them such as accident investigation and reporting is very much a positive aspect of complying with the Act. When one of your colleagues is injured, or if one should die, the traumatic effect it will have on their family and friends will be painful to bear. For you and your business it will be a cloud to live under.

Being safe, and being organised to be safe, means being healthy. This simply requires Health and Safety to be an active key activity in planning and implementing your projects. Planning the procurement and delivery of safe designs, materials and plant into your project takes a little up front time, but ensures assured delivery of your project goals. It needs imagination to judge what a risk is and to what degree there is risk. Thinking about the information available, you can then define a safe method of work. Every thought, every subject considered and addressed, is worthwhile to see your colleagues go home safe and healthy at the end of another busy productive day that was healthy for them, for their business and for your project.

It is necessary for people to go to work in order to get your project delivered, and for them to earn income to finance their lives. In turn they deserve to go home, in as good a state of health as they set out for their work that day.

It's your choice. But it could be their life!

Summary for Health and Safety -

It's always worth your while to make Health and Safety an active key activity in planning, budgeting and implementing your projects.

It's your choice.

But it could be their life!

People

Your project needs people. From clients to engineers; designers to cleaners; accountants to managers; this of necessity would be a very long list if I tried to record everyone – but I can't.

In working with, through and around, the many people that each of your projects needs, you are going to encounter many aspects of people management - motivation, behaviours, actions, reactions, attitudes, likes, dislikes, moods, family situations, births, deaths, marriages, changes, styles, wants, as you develop and deliver your project.

You want to be a project manager, but maybe the role needs a psychologist? To move your project forward successfully you will need to connect up with and relate to very many people. To achieve success you need to be able to motivate, inspire and lead them all. You will get the best from them by enthusing them to focus on your project tasks. In doing so, you will encounter a huge spectrum of human emotions and behaviours. Engaging the support and enthusiasm of people brings ideas, creativity and solutions that you can use to your advantage.

You may like to imagine that everything, once allocated and delegated, is everybody else's problem. In fact, your actions, and inactions, impact on those around you. Everyone is different. Their understanding of issues and resulting actions needs to be clear for them to deliver the needs of your project – but they can easily drift off target. Putting your focus into clarity of goal, and of any intermediate requirement, while enabling each different person to work it out their way, brings effective solutions. Depending on your character this could be the most enjoyable, fascinating and rewarding aspect of project management. Or it may be a trudge – because you are different too.

People can make your project. People can break your project. Working with them is essential. By making your role enjoyable, you will inspire yourself. While you are inspired, you can inspire those working for you and with you to successfully deliver your project.

Giving feedback provides learning and the opportunity to grow in positive ways. Positive feedback provides fuel. Negative feedback, though needed to show where not to go, needs great care – it can so easily wipe out all the good achievement.

Summary for People -

Your project needs people.

To achieve success you need to be able to motivate, inspire and lead them all.

Positive feedback provides fuel.

Conclusion

Why Project Manage? What will all the energy and cost of project managing achieve?

Managing projects is about delivering change. Change as needed by your client. Your client may be clear in every detail in defining what they want. Or they may have just a goal in mind with little view of the involved detail.

Your projects will involve planning for known's, and for unknowns in the form of risks. Some of these risks will most likely mature. When they do they may not materialise in the way that was anticipated, or when you assessed that they may.

Variations from your plan will need to be responded to in real time and managed by you to close the gap and bring your project back on track.

There will be processes that you need to use, and you may need to innovate and create new processes to enable your project to succeed.

Creating your plan will enable you to see how everything fits together, and when each action needs to take place. But you can't spend your time just planning; you need to be doing as well. Monitoring your plan, changing it to keep it up to date is all part of managing the active project. Keeping it simple will help you keep it up to date.

You will need to budget your project based on the plan you have. Payments will need to be made in timely fashion to keep your suppliers on board. Keeping them committed to delivering when you need them to.

Just when you think you are in control, you will find that others are not doing what you agreed with them. Your response will need to be calm, focused and inspirational to everyone across your wider team.

As the components of your project progress through design you will want to drive economic carbon reduction into place to keep costs in check while protecting your planet. Looking after the environment sustains you and your clients business today, and into the future.

To keep the team on track you need to put chasing activities into place, or other projects for other project managers may overtake your priorities leaving your project floundering.

The challenges you face will seem at times like mountains in front of you. But by taking effective action you can remove those mountains, one rock at a time.

As challenges surface and risks mature, you will need to juggle the time, cost quality triangle in identifying solutions and making recommendations to your client or project board.

Working in teams that may be internal to your business, and external from it, will maximise the potential for your projects to succeed. There will be business issues in each of these teams as well as your project issues for you to overcome.

People in these teams are normally good people that want to do a good job for you. They will want to learn, and grow. You need to tell them what they need to know, and show them what they need to understand. Such effort will be rewarded many times over as these individuals develop and increase in ability.

You should find yourself driving forward a clear commitment on health and safety so that everyone understands that they deserve to go home, in as good a state of health as they set out for work that day.

It's likely that you will encounter many aspects of people management. To achieve success you need to be able to motivate, inspire and lead all of the people in your team.

Through all of this experience you will deliver your projects by inspiring those working for you to successfully deliver your project.

Project managing delivers positive change for your client, while changing you and those working for you for the better.

It's thanks to your clear direction, positive support and motivational approach that your client's project will be successful. That is what the energy and cost of project management achieves.

References

[1] Dhillon, B. S. 2013. *Computer System Reliability: Safety and Usability.* Boca Raton, FL.. CRC Press.

[2] National Institute of Standards and Technology (NIST). 2002. News release on the web available at: **http://www.abeacha.com/NIST_press_release _bugs_cost.htm**

[3] Landauer, T. 1996. *The trouble with computers: Usefulness, Usability and Productivity.* Boston: Massachusetts. Institute of Technology Press.

[4] Brooks, C. G., Grimwood, J. M. and Swenson, L. S. 1979. *Chariots for Apollo: A History of Manned Lunar Spacecraft.* Washington, D.C.. NASA Special Publication-4205.

[5] Wikipedia, 2015. *Intergovernmental Panel on Climate Change.* [Online] Available at **http://en.wikipedia.org/wiki/Intergovernment al_Panel_on_Climate_Change** [Accessed 4 April 2015].

[6] Lundin, S.C., Christensen, J. and Harry, P. 2002. *Fish Tales.* London. Hodder & Staughton. More information can be found at **www.fishphilosophy.com**

[7] Johnson, Dr. S.1998. *Who moved my Cheese.* Croydon. Vermilion.

[8] Wilson, P. 1996. *Little Book of Calm.* London. Penguin group.

[9] Lyubomirsky, S., King, L. And Diener, E. 2005. *The Benefits of Frequent Positive Affect: Does Happiness Lead to Success?* [Online]
http://sonjalyubomirsky.com/wp-content/themes/sonjalyubomirsky/papers/LKD 2005.pdf

www.ingramcontent.com/pod-product-compliance
Lightning Source LLC
Chambersburg PA
CBHW070846180526
45168CB00002B/967